"十三五"国家重点研发计划课题（2016YFC0502903）资助
"高等学校学科创新引智计划"（B13011）资助

从书吧出发

回归真实世界

Restoring Reality
Departure from Book Cafe

左 进 著

江苏凤凰文艺出版社
JIANGSU PHOENIX LITERATURE AND
ART PUBLISHING, LTD

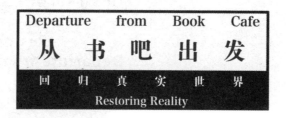

"记录属于 2013 年的我们,一次不同的规划之旅。"

左进 Zuo Jin　　郭泰宏 Guo Taihong　　李晨 Li Chen

吴嘉琦 Wu Jiaqi　　汪梦媛 Wang Mengyuan　　王冠仪 Wang Guanyi

牟彤 Mu Tong　　蒋瑞 Jiang Rui　　刘冲 Liu Chong

林澳 Lin Ao　　林玉 Lin Yu　　杨力行 Yang Lixing　　王卓 Wang Zhuo

序
Foreword

空间的"场所"意念（ideas）经过消费
产生身份"认同感"，成为人们消费的内容

20世纪是工业的时代，人们的消费内容主要为劳动生产的商品（物品），21世纪则是体验经济的时代，人们的思维与行为的价值亦逐渐由工作主义（Work Ethic）转向玩乐主义（Play Ethic），消费内容以玩乐与文化资源相关的商品为主。书与咖啡的消费行为，包括了类似宗教仪式、艺术活动、节庆、社会运动，以及聚会等相关活动的产品，阅读与品味成为空间的媒介(Media)、生产与消费的文化核心。

人们在建筑空间里消费的并非只是咖啡、美食或欣赏艺术品、空间设施等，书与咖啡的消费体验是带有某种"怀旧、文明与流行"的美好想象或深层的回忆情境。不同背景的工作者、建筑师、居民、游客对于"书与咖啡的建筑空间"观看方式的改变，艺术（美学）颠覆传统美学或休闲消费行为，成为一种"社会化"的过程。

左进副教授是一位年轻好学的学者及城乡规划的实务工作者，多年前与我相识，去年左进副教授作为访问学者由天津来我校，我们之间又有了更多讨论交流的机会。"书·吧"空间概念的创造工作，是对持续进行的（次）文化领域的探索，作者利用设计流程与市场调查工作的8个步骤定位建筑功能，再以艺术的手法展示小众消费文化新的主题，丰富并创造休闲性的文化产业。从"书与咖啡的建筑空间"中发现艺术观赏与文化消费，是另一种生活体验方式，是一个体验性的产业。

建筑空间的艺术、文化消费，社交是通过顾客在空间的体验来呈现其内在自由欢愉的休闲意境与游憩体验效益的。本书的出版是他近年来教学新尝试与实践的成果，记录了他努力的历程，因时也展现了他的理想。愿以此序表达相识以来对他的肯定与祝福。

虎尾科技大学休闲游憩系教授兼文理学院院长　侯锦雄
2017年5月30日

目 录
Contents

Hey, 书吧, 你想成为什么?
What do you want to be？

社区书吧
Community Book-cafe

19
社区书吧选址调研
Community Book-cafe Site Analysis

27
儿童中心
Book+Children Center

39
走走停停
Walking in the Bookstore

49
新世纪青少年脱宅中心
A.C.G Children Fun Club

商业书吧
Commercial Book-cafe

59
商业书吧选址调研
Commercial Book-cafe Site Analysis

69
格儿书行
Ger Books

79
春田花花谈书会
Lonely Booktalk

89
冲冲滴手工书坊
Chong's Handcraft Bookshop

后记
Postscript

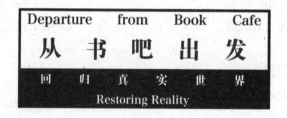

Hey,书吧，你想成为什么？
<Book-cafe>
What do you want to be？

01 02 03 04 05 06 07 08

左进 Zuo Jin | 郭泰宏 Guo Taihong

Hey,书吧,
你想成为什么?

01 项目定位 / 我们要做什么?
PROJECT LOCATION
What are we going to do?

02 核心卖点 / 产品是什么?
SELLING POINT
What are the products?

03 目标客户群 / 他们有怎样的消费习惯?
TARGET CUSTOMERS
What are their habits?

04 市场总量评估 / 这个市场有多大?
MARKET ASSESSMENT
How big is the market?

项目策划要点步骤解析
Projects Planning Analysis

Hey, 书吧, 你想成为什么?

• 功能构成 / 面积比例等
FUNCTION COMPOSING
What are their proportion？

• 成本 / 前期投资成本和后期运营成本
COST
Pre-investment & post-operation

05 06 07 08

• 商业模式 /
盈利模式、运营模式、合作模式

COMMERCIAL MODEL
Profit model, management model, cooperation model

• 选址调研 / 各有利弊
SITE ANALYSIS
What are their advantages and disadvantages？

项目策划要点步骤解析
Projects Planning Analysis

01 项目定位 / 我们要做什么？

PROJECT LOCATION
What are we going to do?

书吧，有很多种可能性。从一开始就要问自己，做什么样的书吧？是针对年轻人还是老年人的书吧？是与电商竞争还是与之互补的书吧？做好定位并让它成为全程策划的出发点和回归点。

Book-cafe has many possibilities. From the beginning, we should determine what kind of book-cafe we are going to run, the book-cafe is for young people or for old people, to compete with E-commerce or complement each other? This is what we will hold onto through out the whole progress of plan.

不同定位的书店的"新陈代谢"

2011年光合作用书店北京两点关闭

2013年4月28日广州学而优书店北京路店关闭

2014年2月下旬季风书园上海华师大店关闭

2014年10月25日西西弗书店深圳cocopark店开业

2015年5月15日方所书店重庆店开业

2014年7月19日单向街空间北京花家地店开业

核心卖点 / 产品是什么？
SELLING POINT
What are the products?

可以说在商业竞争如此激烈，互联网浪潮引起的创新产业蓬勃发展之时，想要有足够的核心竞争力就一定少不了优秀的核心卖点和产品。项目策划第二步就是为商业项目选定一个与众不同的核心产品。

It is reasonable to believe that, with the increasingly fierce competition of business and unstoppable development of creative industries based on the Internet, brilliant main products and selling points are crucial to have adequate competitive strength. The second step for a commercial plan is to discover a unique main product.

目标客户群 / 他们有怎样的消费习惯？
TARGET CUSTOMERS
What are their habits?

目标客户群，是项目最后落成能否活下去的重要因素。目标客户群的调研绝不仅仅是关于人群的定义，更是对人群数量、人群的消费能力、人群的忠诚度进行综合、全面的了解。

Target customers is an important factor decisive of whether a project can survive. Researches concerning target customers are not just about the definition of the crowd, but also about the number of people, their consuming ability, and their degree of loyalty.

04 市场总量评估 / 这个市场有多大?

MARKET ASSESSMENT
How big is the market?

关于市场总量的评估也直接关乎项目的命脉，一个项目规模有多大、给予多大的投资，这些都与市场总量有着极其紧密的联系。对于市场总量的估计需要大量数据的支持，是项目策划定量的重要部分。

The evaluation of the total market is also consequential to the project. How big the project is and how much should be invested are closely linked to the total market. The evaluation of the total market calls for a large amount of data, which is an important part of project quantitative planning.

01 统计调查法
02 预测模型推算法
03 占比加权法
04 德尔菲预测法
05 类比法
06 时间序列预测法
07 市场总量评估方法

功能构成 / 面积比例等

FUNCTION COMPOSING
What are their proportion?

如果说市场定位决定了你的功能设定,那么只有其中每个功能构成良好的搭配,才能使你的市场定位不白费功夫。在项目里不同功能之间如何联系,如何协调它们之间的面积比例、交通关系,都将决定项目的成败。

If the market positioning determines the function, only function structure with good collocation can make the market position worthwhile. How to balance different functions and thus decide the ratio of area allowed to each, and the internal traffic, as a whole, will make or break a project.

案例分享——库布里克书店

电影纪念品区
原版电影
电影海报、电影影碟
电影原版书籍

设计产品区
设计书籍
香港本地制作 / 海外手作
精品杂志

特色功能
图书发行、音乐会
驻场作家、创作人工坊
诗会、一人一故事剧场

06 选址调研 / 各有利弊

SITE ANALYSIS
What are their advantages and disadvantages?

李嘉诚先生说过商业最重要的要素就是地段，一个项目的区位和这个区位所带来的成本、文化和风情，将会直接决定这个项目的发展之路。

As Li ka-shing, the richest man in HK, once said, the single most important factor in a project is the location. Location and, coming along, the cost, the culture and customs, will directly decide the development of this project.

案例分享——沙坡尾艺术西区

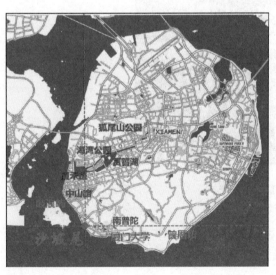

厦门沙坡尾

位于鼓浪屿与厦门大学之间，作为废弃的避风坞一度环境很差，但却也因此成为价值洼地。与高端商业区相比，有不同的成本和格调，因此沙坡尾改造计划实施后，有大量的青年潮男涌入，整个地块遍布了各种各样消费水平较低、文化程度较高的业态。

成本 / 前期投资成本和后期运营成本

COST
Pre-investment & post-operation

项目的成本关乎项目的存活，如何在获得最大利润的情况下，投入最少的成本是一个成功的经营者的最高追求。项目成本既包括了前期的建设成本，更包括了后期宣传、原材料、人事等许多直接和间接成本。

The cost of a project is related to the inventory. How to obtain maximum profit at least cost is the highest pursuit of a successful operator. Cost includes not only direct cost of construction in starting period, but indirect cost such as publicity, raw materials, etc.

案例分享——"巢"书店的营业模式

在沙坡尾有一家名叫"巢"的小书吧，店铺很有意思，老板大多数情况下都不在店内，店内提供咖啡饮品、读书阅览和娱乐聚会等功能，消费者在买单时完全自助，按照自己的消费感受来付款，金额不限。可以看出在这个经营模式下，老板节省了大量的人力成本，同时由于选地在一块不折不扣的价值洼地，建筑也只是进行了简单改造，低廉的成本为店铺的存活提供了有利条件。

商业模式/盈利模式、运营模式、合作模式

COMMERCIAL MODEL

Profit model, management model, cooperation model

项目的产品、项目的用户、项目的市场、项目的盈利模式,这些共同构成了项目的商业模式。一个好的项目一定是基于一个十分优秀的商业模式,这里既包括了如何宣传,又包括了如何合作,更包括了如何运营。整个项目只有在一个有机整体的商业模式下才能成功。

The commercial model of a project consists of products, users, market and profit model. A good project must be based on an excellent commercial model, including how to publicize, how to cooperate, how to operate. The whole project can only succeed in a very organic commercial model.

商业模式内容

产品模式
你提供的产品是什么?能为用户创造什么样的价值?你的产品解决了哪一类用户的什么问题?

用户模式
作为创业公司,你一定要找到对你的产品需求最强烈的目标用户。

市场模式
真正的推广模式是要根据你的用户群和产品,去设计相应的推广方法。

盈利模式
收入模式,就是在通过产品获得巨大用户基数,在此前提下考虑怎样来获取收入。

book-cafe

商业模式不是赚钱模式，它至少包含了四方面内容：产品模式、用户模式、推广模式，最后才是收入模式，是指怎么去赚钱。

一句话，商业模式是你能提供一个什么样的产品、给什么样的用户创造什么样的价值，在创造用户价值的过程中，用什么样的方法获得商业价值。

如今的商业模式更加注重推广，推广之后便会出现反馈，这个时候如果不对产品进行调整，你和团队将面临非常大的挑战。真正的推广是对产品的不断完善和提升。在推广的过程中，你要研究市场，跟目标用户打交道，了解用户真正的需求，了解用户使用产品时遇到的困惑和问题，再反馈到产品上进行改进，由此不断调整和完善。

Today's commercial model centers more on promotion, followed an instant feedback. If erroneous products are not adjusted, you and team will face a very big challenge. The real promotion is to constantly improve and upgrade the product. In the process of promotion, you need to research the market, dealing with target users to understand their real needs, confusion and problems they encountered when using the product, and then improve the product, thus constantly adjusting and making perfect.

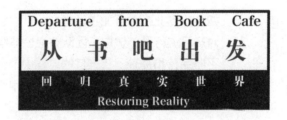

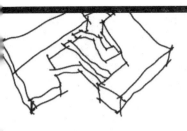

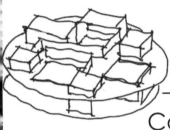

社区书吧选址调研
Community Book-cafe Site Analysis

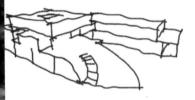

儿童中心
Book+ Children Center

走走停停
Walking in the Bookstore

新世纪青少年脱宅中心
A.C.G Children Fun Club

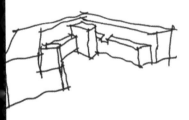

王冠仪 Wang Guanyi　　蒋瑞 Jiang Rui　　杨力行 Yang Lixing
林玉 Lin Yu　　牟彤 Mu Tong

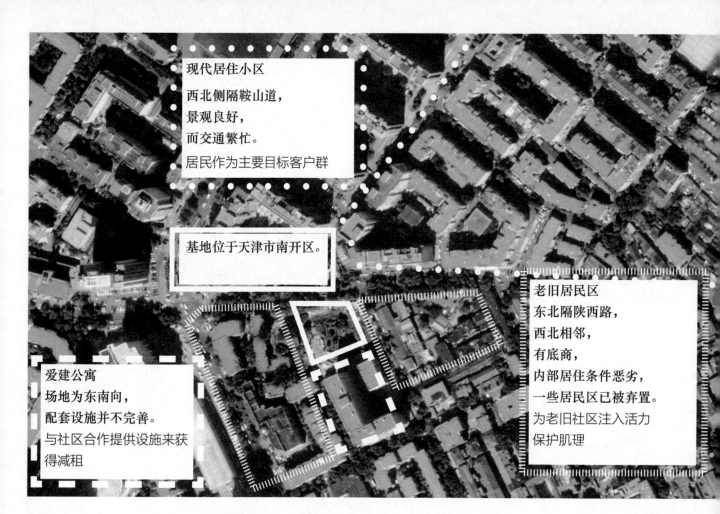

现代居住小区

西北侧隔鞍山道，
景观良好，
而交通繁忙。
居民作为主要目标客户群

基地位于天津市南开区。

老旧居民区
东北隔陕西路，
西北相邻，
有底商，
内部居住条件恶劣，
一些居民区已被弃置。
为老旧社区注入活力
保护肌理

爱建公寓
场地为东南向，
配套设施并不完善。
与社区合作提供设施来获得减租

There is a modern residential district across Anshan Street on the north-west, but there is masses of traffic.

Residents there can be served as main customers of our project.

The site is located in Tianjin Nankai district.

Aijian apartment is in the south-east of our site, facilities there is imperfect.

We can cooperate with the community by offering facilities and gain rental discount.

An old residential district across Shannxi Road, adjacent to northwest. Some shops on the 1F. living condition is bad, some of them are discarded.

Add vitality to the old residential district and protect the texture

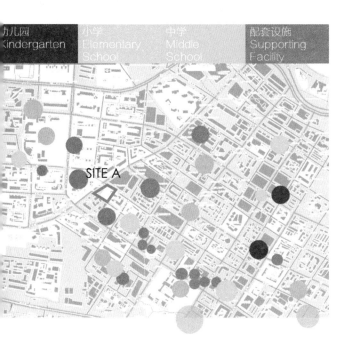

社区性
Community

区域内教育设施规模
Scale of educational insititute in the neighborhood

通过了解整片区域，我们认为，应该设计一个服务性的书吧。因为周围多是居住用地，生活气息浓郁，且有丰富的老旧建筑营造出的历史氛围。所谓服务性也是社区性，需要从结合当地居民需求的角度去构思我们的项目。

As we have fully recognised the area, we came into agreement that we should design a service-inclined book-cafe, because residential land occupy most area of the region and the historic building are bearing memories of community. Service-inclined also means community-inclined, so we need to think of the need of residents when designing the project.

为什么选择青少年？
Why Teenager?

儿童感知世界的方式决定了他们对"实体"有更强的依赖性，图书在"亲子阅读"中充当桥梁的作用，具有不可代替性。

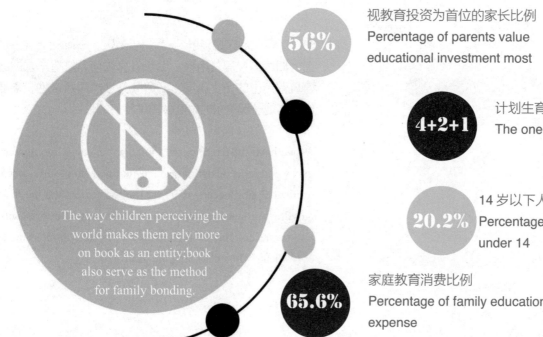

The way children perceiving the world makes them rely more on book as an entity; book also serve as the method for family bonding.

56% 视教育投资为首位的家长比例
Percentage of parents value educational investment most

4+2+1 计划生育模式
The one-child policy

20.2% 14岁以下人口占总人口比例
Percentage of population of people under 14

65.6% 家庭教育消费比例
Percentage of family educational expense

通过调研区域内的公共服务设施，我们发现周围有足量的文化教育设施，这说明区域范围内有足量的儿童，但缺乏儿童和青少年的娱乐休闲场所。为什么我们不能做一个服务性的儿童书吧呢？服务儿童必然也随之服务了他的家庭。

由上述教育学的观点和中国政策形式可以看出，书籍作为一种文化知识的载体，更易让家长接受而购买。面对如此庞大的市场金矿，各个出版社纷纷踏入其中，争取分得一块蛋糕，形成目前中国儿童图书市场异常火爆的景象。

When investigating the area, we find that there is sufficient educational institutes for children, which indicates enough number of children. But there is not enough entertaining facilities for them. Why not design a service-inclined book-cafe for children? Serving children is along with serving their family.

According to evidence above, book, as a carrier of culture and knowledge, is more attractive to parents. Facing such a juicy market, publishers are all into earn the profit, which makes a hot children book market nowadays.

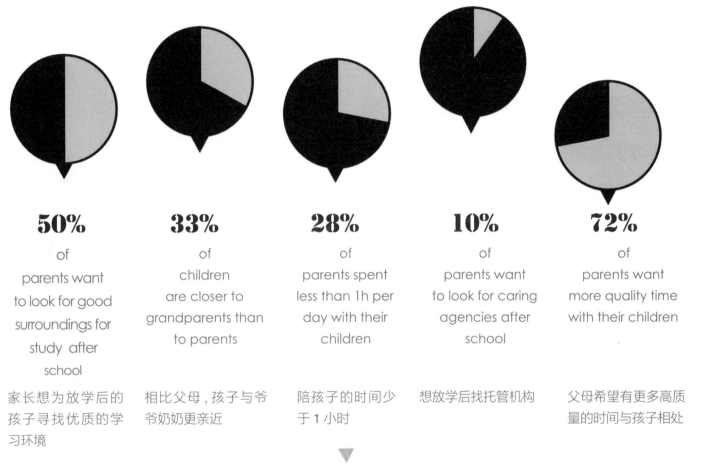

"亲子"为我们带来了什么？
What can family-bonding do for us？

由"亲子"这一主题我们不难想到一个需求量很大的服务项目——托管。

It is not hard to combined family-bonding with children carring.

托管班是由公民个人依托非国有资产举办的机构

由企事业单位、社会团体和其他社会力量举办

其直接领导单位是各级民政局

法律体系不完善 — Relevant law is not prepared

社区托管中心处于"无政府状态" — Children care centre are not supervised by any government

健全相关法律呼声较高 — People are appealing to complete relevant law

区域内托管市场前景广阔，竞争较少。由此我们得出结论托管可以作为维持项目生存的手段，但要优先建立行业标准，良心经营，以获得社区和政府的支持。

The children caring market is juicy and there is little competition. So we can make the conclusion that children caring is the measure to keep the project survive. However, we must hurry to set standards for others in order to gain support of the community and government.

了明确的目标后，我们针对附近的学校和儿童的需求进行更深层次的调研以辅助我们进一步的设计。

餐饮是大家认可的经营项目之一，因此我们协同调研周围餐饮业状况和需求。

With a definite goal, we investigate more deeply schools around nd needs of children to help to develop the design.

Catering is one of project acceptable by us, thus we investigate the catering situation and needs around by the way.

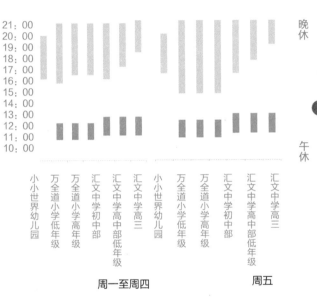

区域内中小学生休息时间段

周一至周四　　周五

调查附近学校上下学的时间可以帮助我们分时段运营。
nvestigating the school time helps us to run the book-cafe ccording to time table.

不同时段针对不同人群

营业时间 11: 00 - 20: 00（工作日）
DPENTIME 9: 00 - 21: 00（周末、节假日）

餐饮
临时托管+亲子活动

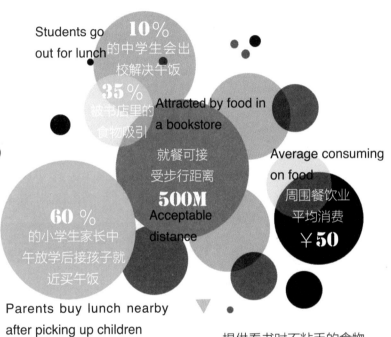

Students go out for lunch 10% 的中学生会出校解决午饭

35% 被书店里的食物吸引 Attracted by food in a bookstore

就餐可接受步行距离 500M Acceptable distance

Average consuming on food 周围餐饮业平均消费 ￥50

60% 的小学生家长中午放学后接孩子就近买午饭
Parents buy lunch nearby after picking up children

提供看书时不粘手的食物
Provide tack-free snacks when reading book

家长等候孩子时的饮料
Beverages when parents are waiting

亲子 DIY 厨房
Diy kitchen

适合中等消费人群
Affordable for moderate consumers

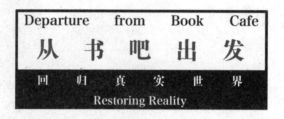

Departure from Book Cafe
从 书 吧 出 发
回 归 真 实 世 界
Restoring Reality

儿童中心

Book+ Children Center

王冠仪

Wang Guanyi

"爱画画，运气不会差。"

项目定位
Project Location

儿童自主阅读
Independent reading

家长温馨陪读
Family reading

伴读员科学伴读
Accompanied Reading

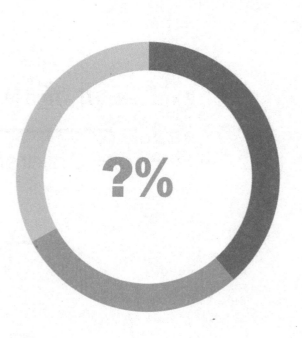

该儿童中心是一个以私人定制式伴读为主要产品，以为孩子提供发展个性的活动场所为理念，以创造温馨的亲子空间为亮点的项目。其主要产品是伴读，家庭可以根据不同的情况分配时间进行不同类型的阅读。

BOOK+ children center's main product is specially-designed accompanied reading. Our pursuit is to provide children with a platform to develop their potential and our specialty is to create a family-bonding space.

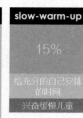

核心产品
Main Product

Customers can use products at their will, to ensure the quality of online shopping

We provide place and facilities for children

营养餐饮
Food

儿童乐园
Children resort
提供游戏器械和场地

儿童精品
Children necessity
儿童中心网页上的童装文具、玩具、书籍、即食餐饮，均可在此体验并由家长在网上自选套餐

Public competition will be held to provide chance to pormote the family

Reading gathering and education lectures will be held from time to time

社会交流
Social connection
天使宝宝明星妈妈大赛，提供与社会分享生活乐趣的机会

亲子活动
Family-bonding
不定时举办亲子阅读会、育儿讲座等活动

伴读
Accompanied reading
专员接放学的孩子到儿童中心后，可以由经培训的伴读员陪伴他们读书，也可自行安排，由家长私人定制

职业体验
Job-experiencing
为孩子创造体验社会的机会，帮助孩子更好地发展个性，有别于传统的技能培训班

Class which need cooperation of parents and children

亲子课堂
Family class
设置一些需要孩子与父母合作完成的任务，加强他们的沟通和相互了解，比如亲子厨房

摄影集
Photo

We can record lives of children for the family

Provide children with chances to experience the society

儿童中心提供伴读、社会实践、儿童精品等服务项目。会员家庭可以通过服务券来兑现服务，服务券可以购买，或通过提供活动场所或照顾儿童来获得。

Book + Children center provides accompanied reading, social practice, children's boutique. Members can order services in change of "SERVICE COUPON". It can be bought or obtained by providing activity place or looking after children.

CHILDREN CENTER
BOOK+

合作模式
Cooperation Model

BOOK+ 采用线上线下结合平台式经营模式，与社区、教育机构、供应商合作，提供精细的、多样的产品，并减少经营成本。

We are selling products both online and offline and also cooperate with communities, educational institutions, supplier to provide elaborate and various products, reduce operating cost.

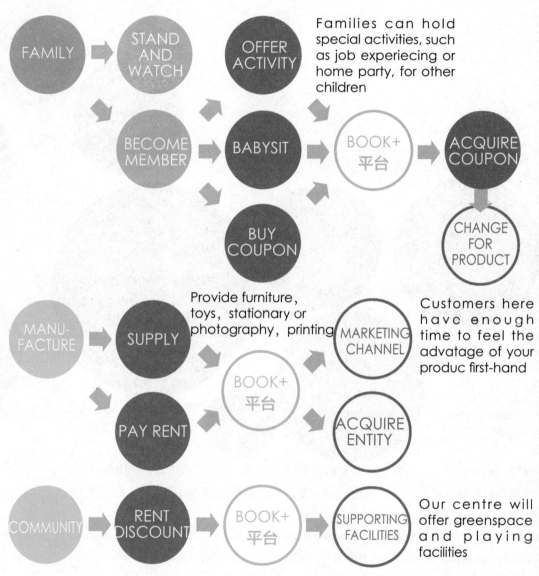

Families can hold special activities, such as job experiecing or home party, for other children

Provide furniture, toys, stationary or photography, printing

Customers here have enough time to feel the advatage of your produc first-hand

Our centre will offer greenspace and playing facilities

成本估算
Cost Estimation

过估算单位面积产出来确定功能面积比，以现利益最大化。

By estimating the output per unit to confirm the functional area ratio, then to maximize the profit.

利润源 SOURCE
儿童家长、儿童产业、社区
Parents, Industry, Community

利润点 PROFIT
托管费、课程培训费、比赛报名费；租金、佣金；餐饮、活动门票
Children-caring, classes, competition entry fee, rent, commission, tickets of catering and activity.

利润杠杆 LEVERAGE
雇员审核监督、网络运营、转租店铺的监管、装修
Employee Supervision, network running, supervision and decoration for subleased shops.

利润屏障 BARRIER
网络技术发展速度飞快，所谓利润屏障很难形成，唯一的办法就是发挥前瞻性，快速升级
Network technology develops fast, it's hard to form profit barrier, the only way is far-sighting, fast-upgrading.

成本估算 Cost

项目 Item	费用 Expense
家具	15 000 元
软装	800 元 / m²
草坪	9 元 / m²
儿童乐园设施	500 ~ 1000 元 / m²
用地租金	50 元 / (m² · 天)
伴读员、少儿教师等兼职人员	75 元 / (h · 人)
服务员	100/ (天 · 人)
食材、学生营养餐配送	15/ 人次
电费	100 元 / 天
水费	0.15 元 / (儿童 · 日)
设备维护和网络运营	

单位面积产出 Output per unit area

区域 Area	收入 Income	人均面积 Average Space	成本 Cost	单位面积产出 【元 / (h · m²) 】 Output
公共阅览区		5 m² / 人		吸引潜在客户群
托管区	25 元 / (天 · 人)	7.5 m² / 人	5 元 / (天 · 人)	2.67
教室	45 元 / h	15 m² / 儿童	10 元 /h	2.33
转租店铺	100 元 / (m² · 月) +5% 佣金			0.3+
餐饮	25 元 / 人	3 m² / 人	15 元 / 人次	6.67
儿童精品体验	3% 佣金			产出包含在阅览区内
儿童乐园	20/ 人			
室外场地租金	75 元 / (天 · m²)			2.3
网络平台使用	650 元 / 月 +5% 佣金			

项目能够提供一种适合亲子阅读的空间，并为临近街区注入活力，服务社区。为了提供适合家庭的亲近尺度并满足儿童心理的探索需求，建筑采用集装箱这一极具灵活性的建筑形式，它不仅可以通过组合营造多种多样的空间，也利于结合网络监控技术来实现灵活分割，便于运营和保障儿童安全，还有利于未来发展和扩建。

建筑主要分为两个体量，主要的儿童阅读区域在南侧，其中每个集装箱作为单个家庭的活动单元，3～4个集装箱组合出半开放的过渡空间，最大的公共阅读区域在其中起统筹作用。其余辅助功能和社区服务功能在北侧。

The project can provide space suitable for parent-child readin add vitality to the adjacent street, serve community. In ord to provide an intimate space for family and varied distanc for children to explore, the project uses container as its ma construction material. It not only can provide space with ri variation by different combination, but facilitate management different section with help of network monitoring technology.

The architecture mainly consists two blocks. Reading area f children is in the south. Each container serving one family, 3 4 containers form a semi-open area as transition space and t largest open reading area integrate all. Other auxiliary functio and area for the community are in the north.

建筑设计
Architecture Design

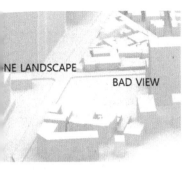

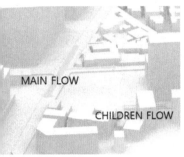

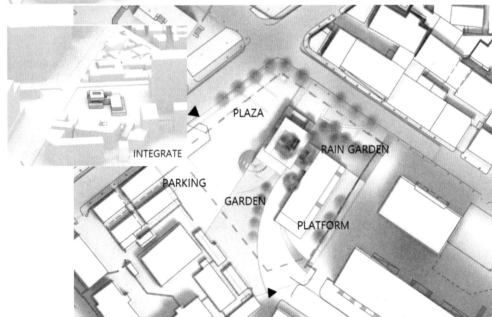

为了更好地服务社区并营造良好的街道景观，建筑周边有大面积的公共绿地作为发展用地，建筑内引入了丰富的植物景观，中庭的绿化与垂直交通相结合起到连接两个体量的作用。

In order to better serve community and create good street landscape, there is large public greenbelt around the architecture. As developing land, the inner architecture introduces various vegetarian. Planting in the atrium combined with vertical transportation serve to link the two blocks.

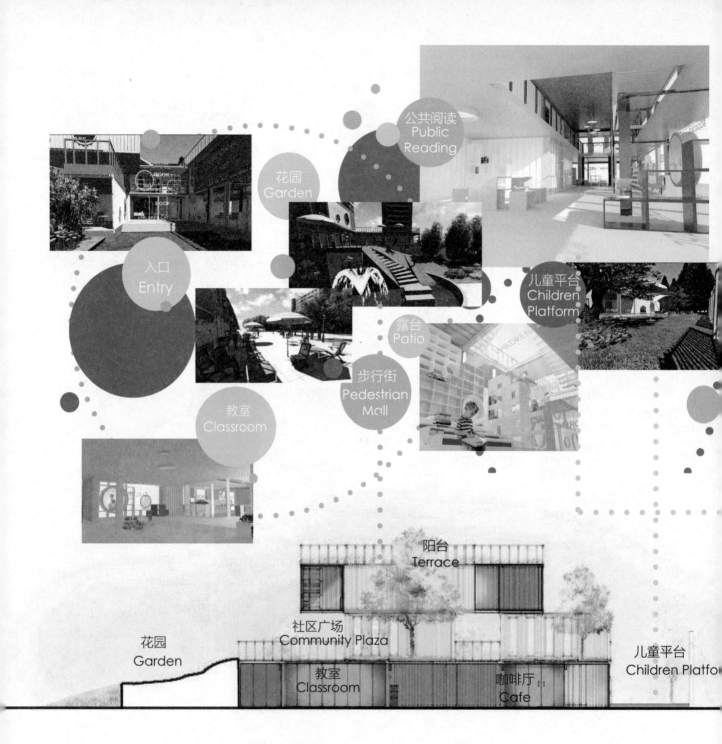

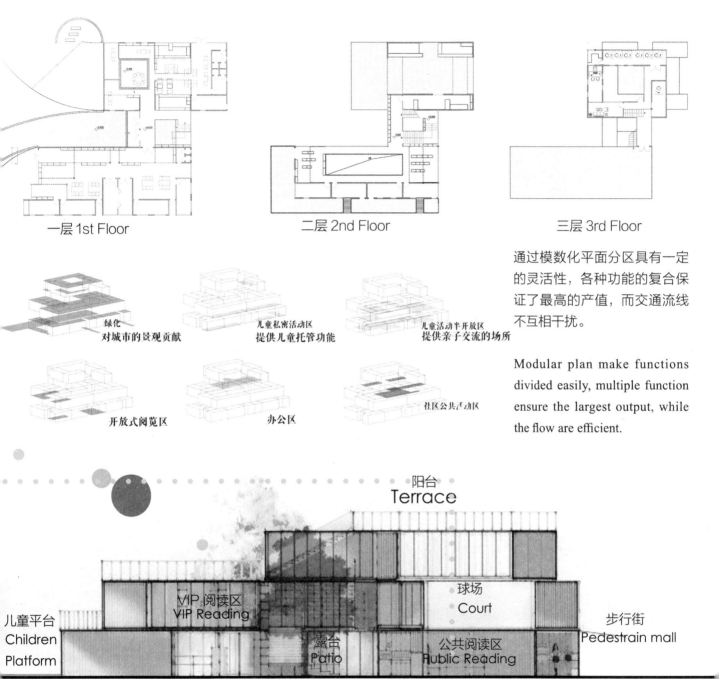

分时运营
Open Time

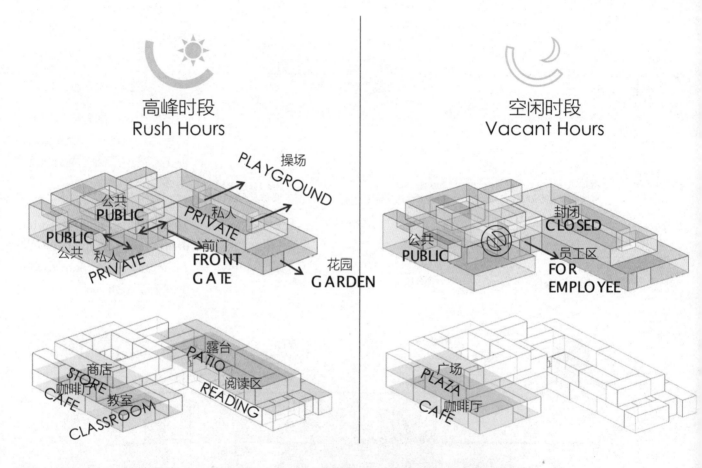

根据之前的调研分不同时段运营,通过家具组合改变空间的功能,并靠 NCF 技术实现管理。

According to the previous research, it will run at different times. We change the combination of furniture and manage with the help of NCF.

"**实**体书店变得难以生存""儿童书吧的需求在增长",
如何权衡这两者是我思考的切入点:
3000 m² 的面积里需要有什么功能才能满足需求?
怎样才能让先锋建筑在这样的社区氛围浓厚的基地上扎根?
以书为关键点的儿童中心是我的回答。
这个关键点还串联着教育、托管、家庭服务等第三产业,足够与如此庞大的建筑面积相匹配。要让这个建筑"接地气",亲子教育与家长的工作繁忙是一个亟待解决的矛盾,"关爱孩子成长、促进亲子交流"成了这个项目的广告语。

Entity bookstore is diminishing,*Needs for children bookstore is roaring*,how to balance the two above is the key: the area of 3000 square meters can contain what function to meet the needs? How to make the pioneering architecture take root in such a base full of rich community atmosphere?
Children center taking book as the key point is the answer. The key point also connects education, children-caring, family service, etc, which match with such large area. In order to make the architecture functional, it is urgent to solve the problem—parenting and the busy work of their parents, so caring for children, enhancing family bonding become the advertising slogan.

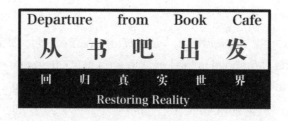

走走停停

Walking in the Bookstore

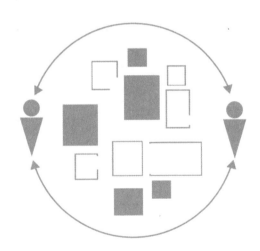

蒋瑞

Jiang Rui

"每一天的到来都万分惊喜。"

从需求出发
Based on the Needs

居民年龄构成以 19～40 岁人群为主,他们对交友、聊天的需求较旺盛,大多数人倾向于和身边的人面对面交流。

Residents mainly consist people aged from 19 to 40, they are in need of chating. Most of them incline to communicate face to face.

居民年龄构成
Age Composition

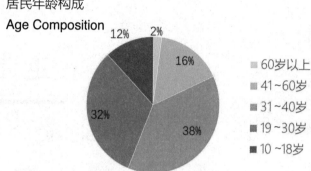

居民娱乐生活需求
Needs of Entertainment

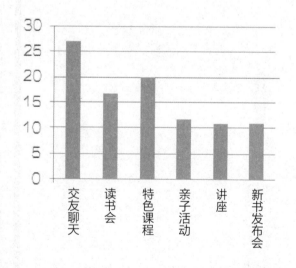

居民渴望的交流形式
Ideal Communication

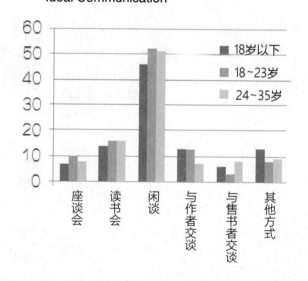

经营模式和产品
Management Model & Products

 +  +  =

线下服务
Offline Service
活动策划与举办
场地提供与租赁
Events planning and hosting
Providing and lending place

线上体验
Online Experience
活动信息推荐
相关产品购买
Promotion event
Selling products

吸引人气
Advertising
就近社区内传播
朋友圈分享机制
Within community
Smartphones

利润
Profit
在资金平衡的前提下致力于社区服务
Servicing community by all means

Product 1 — 一周生活美学 — 50%
饮茶、交友、教育子女、练习琴棋书画
Tea, Socializing, Drawing, Music, Writing

Product 3 — 私人会所 — 25%
日租、时租，提供桌游、书籍
Private Party, Card Games, Reading books

Product 2 — 阅读邻居 — 15%
从邻居那听来的故事、推荐电影
Stories of the Neighbour, Recommending movie

Product 4 — 读书计划 — 10%
线下体验阅读、线上扫描购买
Reading Offline, Purchase Online

社区人剪影
Community Group Profile

林小玉 活泼中学生
——"三好学生好学好玩好聊"
和同桌汪小媛一起自习,为高考努力
和舍友一起去桌游
Lin Xiaoyu, Middle School Students
——"Study well, Play well, chat well"
Self-study with classmates
play games with roommates

林小澳
——"我穷但我爱生活"
享受安静的阅读空间
游走、散步、穿行、发呆
幻想邂逅
观展大开脑洞
Lin Xiaoao,
——"Love life even I'm poor"
Quiet reading space
Wandering, Walking, dazing
Fantasy for encounter
Exhibitions for inspiration

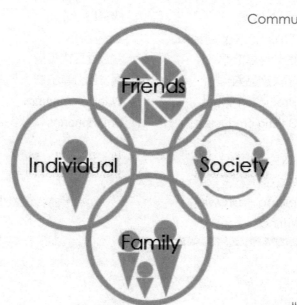

社区人 被微信绑架生活
——"我不想再宅了!!"
在家门口参加有趣的活动
交新朋友
Community Group, indulged in Wechat
——"I hate staying at home"
Participating in funny activities
Making new friends

左帅 好爸爸
——"喜欢陪伴家人"
和家人散步,看宝贝蹦蹦跳跳
给孩子读书讲故事
让宝贝和别的孩子一起玩耍
Zuo Shuai, Good Father
——"All I want is to be with my family"
Walking with family
Seeing baby hopping
Reading with baby
Watching baby play with others

方案生成
Programming

基地周围建筑密集
Compact texture

利用场地连接多条路径，使之成为路径穿梭的地方
Create cross paths

基地居民交流局限于小区内，沟通的机会很少
Communi-cation is limited

沿路径在城市肌理上加上"盒子"
Along paths add box in city texture

居民出行路径分散，缺少人流汇聚点
Travel route is scattered, lack of gathering point

用圆形底座来保持匀质性
Round base to keep homogenity

建筑设计
Architecture Design

小盒子的体量与周围街区相当，整个建筑融入街区，但又不会消失，将人们聚集在一起，创造着新的相遇与相识。

The volume of the box matches with the surrounding blocks integrates into the blocks, meanwhile, it is special and attracting, which creates the chance between people to know each other.

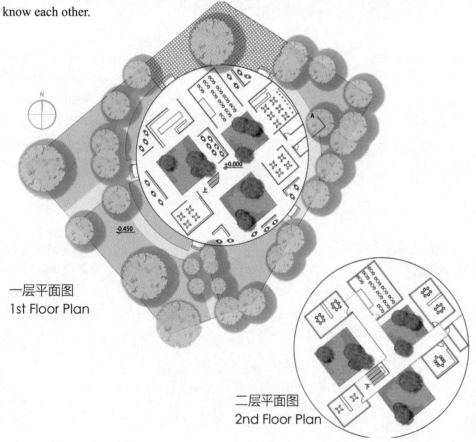

一层平面图
1st Floor Plan

二层平面图
2nd Floor Plan

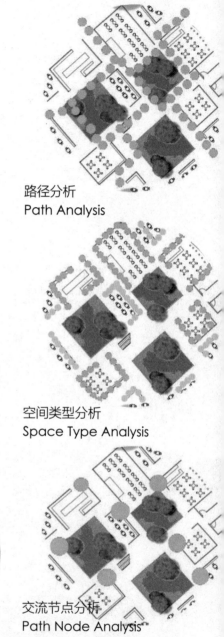

路径分析
Path Analysis

空间类型分析
Space Type Analysis

交流节点分析
Path Node Analysis

是一块充满活力的土地，人们来来往往，在这里生活、工作，开车经过，或是散步路过，一旦建筑建成，便对周围的一切造成影响，变成一股能量，散发出去，而接下来，将和这里的人与事物一起生根，彼此融合。

This is a living site, people passing by, living and working here. Once the building finished, it can make an influence to the surroundings, with kind of energy emitting, and next, the building can get own life with the people here, mixing together.

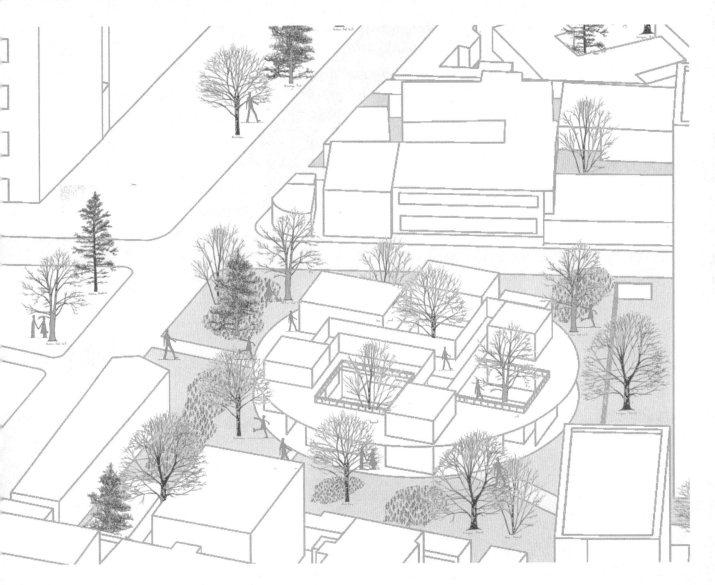

以"走走停停"为设计构想，建造一个可以游走的书吧，一层流动的路径，隔出功能需要的小室，人们可以在此处闲逛游走，看书、喝茶、聊天，而二层则是小室，供三两好友聚于其间，侃天说地。

Taking "walking in the bookstore" as design concept to build a book-cafe, it can provide various ways to relax, chat with others and read books. They can saunter at first floor or talk with their friends at second floor in a room.

这个项目中，询问了当地人的需求，考虑周围的建筑是否与自己的设计异质，然后不断地发问、不断地改进，渐渐形成了自己的方案，这是设计师的责任。

In this project, the designer asks local people's needs, considers whether the surrounding architecture is different from the design. Then the designer keeps asking and improving, forms the design program at last, which is the designer's responsibility.

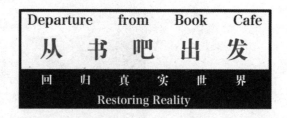

新世纪青少年脱宅中心

A.C.G Children Fun Club

杨力行

Yang Lixing

"人的行为可以懒惰,但是思维不能懒惰。"

做宅
别做死宅

关于"死宅"
About the Walking-dead

现如今，越来越多的青少年沉迷网络、游戏和动漫文化，变成了足不出户的宅男宅女。长期与社会隔绝使得他们变得性格内向、自我封闭，逐渐变成了所谓的"死宅"。

Nowadays, more and more teenagers are indulged in the internet, game and animations, they become "otaku" who rarely go out of house. Long time isolation make them shy and reserved, at last they become the "walking-dead".

做一个健康的"宅人"
Otaku but Positive

而我也是动漫文化爱好者的一份子，因此在接触到大量宅人群体之后我发现了这个群体所存在的问题，因此我要去改变这些人内向、自闭的状态，纠正大众对这个群体的偏见。

As a lover of animation culture, I found the problem of t group. So I am trying to figure out a way to solve it, change t prejudice of the public.

第一话
死宅出门!

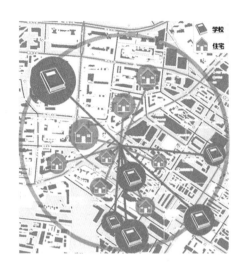

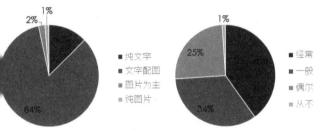

青少年最喜欢的图书类型调查　　青少年阅读漫画的频率

中国漫画市场现状
Current Situation of Chinese Comic Market

画消费通常存在于缺乏娱乐消费的地方。
onsumption of comics usually happens where lack ntertainment substitution.

国漫画市场呈现低龄化趋势，中小学生消费额占比大。
onsumption of comics are mainly made by junior or ementary students.

文字和图片混合类图书最受欢迎，具体类型包括童话、幽默故事、探险故事等，当然少不了漫画。漫画阅读成为青少年的主要休闲娱乐方式之一，其受众群可观。

Books with both graphics and words are the most popular ones, including fairy tale, funny story, adventure story of which comics are typical.
Reading comic books is becoming main entertainment for teenagers. So the market is considerable.

场地现状
Site Condition

场地中中小学集中，存在大量潜在客户群。
There are plenty of junior and elementary schools in the site, so there are a lot of potential customers.

场地中缺乏青少年娱乐设施。
The site is lack of entertainment facilities.

来看
真正的好动漫

第三记

动漫休闲 + 社交场所
Animation+Communication

在设计我的书吧时，我希望我的书吧能够成为吸引宅人一起来聚会的社交平台，从而为这些宅人找到一个能够与他人交流的平台，创造一种积极向上的动漫文化氛围。

I hope that my book-cafe can be a platform for the "otaku" to get together and create a positive atmosphere of animation culture.

一起
来看漫画
精品漫画悦读中心
以漫画阅读为主题

绘制: Tony-tony

来看
真正好动画
免费和大家分享
每晚一部精品动画

绘制: 333

绘制: rio

边看漫画
边学画画
趣味绘画教
课外绘画活动中

A.C.G. Children Fun Club
新世纪青少年脱宅中心

第四话

同学，顺便买点东西呗

- 不求卖出商品只求**瞩目**
- 以商品吸引**客群**
- **线**下与**线上**结合
- 以**低价**商品打头

绘制：doge leg

以精品漫画和绘本阅读为主题，吸引社区青少年在空闲时来此进行休闲阅读。定时举办免费的动画电影放映会，向孩子们推荐优秀的精品动画电影，一方面成为吸引顾客的卖点，一方面通过这种活动引导青少年接受健康高雅的动漫文化形式。和以上两种活动形式相结合，组织绘画和手工教学课程，让孩子在看的过程中提升学的兴趣。

Taking qualified comics and picture books as the subject, the book-cafe attract children to read when they have time. Animation movie will be cast timely, on the one hand it will attract customers, on the other it can improve tastes for animation. Craft class combined with activities above will cultivate children's interest.

第五话
A.C.G.Club 的未来

舒适的阅读空间
Comfortable Reading Space
精品的阅读种类
Specified sort
饮料、桌游……
Drinks and card games

学生午托
Children Noon Care
绘画教学
Teaching drawing
商业售卖（书籍、周边）
Product Selling

图书线下试读
Offline reading
商品线下展览体验
Offline product trying
线上销售
Online selling

咖啡厅外包
Café renting
降低运营成本
Lower cost

吸引对象人群
Attracting Customer

使人驻留
Maintain People

社交平台功能
Communication Platform

商业项目
Business Project

与社区合作
Cooperate with Community

与线上商店合作
Online store

与出版社合作
Cooperation with publisher

与咖啡厅合作
Cooperation with Café

免费漫画阅读
Free comics reading
免费动画放映
Free animation watching
免费周边展览
Free products exhibition
……

绘画手工课程
Handcraft lesson
动漫俱乐部
A.C.G. club
漫友交流会
Acer communication
特色讲座
Special Speech
……

吸引目标人群
Attracting customers
联合社区开展活动
（吸引客户群）
Cooperate with community

进行专题图书宣传
Promote reading
举办特色活动
Holding special events

第六话
故事的最后，
建筑设计

在空间设计上该书吧以中心庭院为核心，希望创造一个能够与外界的嘈杂环境相隔离的自由阅读环境。同时在书吧主题上，希望能够通过该书吧建立一个社区青少年动漫文化活动中心，鼓励喜欢动漫文化的青少年走出家门积极与他人接触。

The project featured by animation is aimed at the teenagers. The space surround the yard, creating an isolated and tranquil environment. I hope to establish an animation centre for the adolescent, and hence encourage them to go out of the door. There will be a positive atmosphere for communication.

交通 Transportation　　入口 Entry　　周边 Neighbour

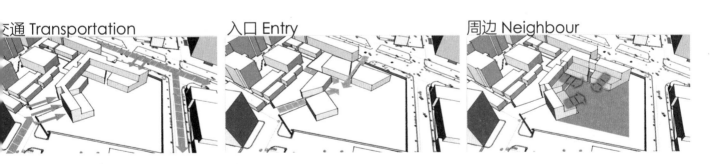

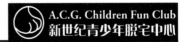

缩小餐饮区面积,增加公共大空间;
一层作为主要公共区,对空间进行合并;
商品展览与阅读区融合。
Cut down the area of dining space, add to public open space.
The first floor serve as main public space, incorporating spaces.
Incorporate display section with reading section.

二层作为私密区,对空间进行细分。
The second floor serve as private section, divide space.

室外空间的多重利用
开放阅读
写生
放映电影
体育活动
Multi-use of outdoor space
Open reading
Sketch from nature
Movie projection
Exercise

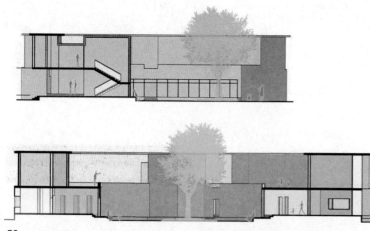

在进行书吧的平面布局设计时,希望书吧建筑能够形成一个内包围的绿地公园,打造书吧室外阅读空间。因此书吧整体呈半包围状将围合空间与外部街道分隔开来,同时在场地的几个主要出入口设置了可以直达内部绿地的通道,希望能够利用中心绿地引导街上的行人进入游览。

As for plan of the book-cafe, the centre is surrounded by the green park and make it an out door reading area. Therefore, the building separates the inside and outside. Meanwhile there are roads leading to the green park and some main entrances, which can attract customers of the street.

项目要求带有目的性和逻辑性去设计，从前期调研到方案设计再到后期运营模式一体化。在后期服务模式的设计中，需要像运营商一样去了解使用者的需求，在调查和迎合使用者需求的同时，挖掘书吧的潜在需求，发现其潜在价值，初步形成了设计思路，为项目的逻辑性和合理性奠定了基础。

"脱宅中心"这个理念是有它的市场的，而且与现今热门的解决"亚健康"的追求相符，设计师的需求调查也证明了这一点。

The project from previous research to the project design even to the late operation model integration needs to be finished with purpose and logic. During the process of the late service model, we need to know the users' needs as what the operators did. When investigating and meeting the users' needs, we exploit the potential needs of the book-café, find the potential value, and form the design idea initially, lay a foundation for the project's logic and rationality.

The concept of this project has its own market, and coincides with the pursuit to solve the hot topic "sub-health", which is proved by the need research done by the designers.

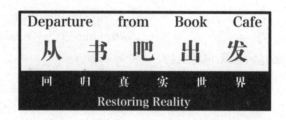

商业书吧选址调研

Commercial Book-cafe Site Analysis

商业书吧
前期调研

在基础性的选址调研中,我们选定的地块既拥有保留十分完整的窄路密网肌理,同时这个区块由于缺乏维护和管理,硬件配套设施并不完善,随着社区环境的恶化,越来越多的原有居民迁出,大批外来人口进入。因此该地块尽管区域位置优越但租金价格普遍不高,是一块不折不扣的价值洼地。

We selected the site which has retained traditional narrow road and dense texture. Meanwhile, facilities there are not complete for lack of maintenance and management. With the deterioration of environment of the community, more and more original residents emigrated and a large number of immigrant population emerged. Despite the advantage of regional position, average rent price in the concession is not high, which means it is a real value area.

该地块的步行系统是它的特色之一,窄路密网给予整个社区便利的可达性。以赤峰道古玩市场为主体的特色商业街是该地块人气来源的一部分。步行小道中有很多文化内涵很高的业态,包括小型的咖啡馆、小型的青年旅舍和小型的花店。

The pedestrian system in this plot is one of its characteristics, and the narrow paths and dense network provide the community convinience and high accessibility. The site, with Chifeng Road Antique Market as the main body, was part of the attraction to the people. There were many businesses with high culture connotation alongside this walking trails, including small cafes, small youth hotels and small flower shops.

选址:锦州道和新华路交口处。现状为一栋老旧住宅,产权状况为公产,由和平区房管站统一管理。周围是新华路电子娱乐街,是许多青年的聚集地,同时距离和平路只有一个街区,区位优势十分明显,能吸引大量人流。

Location: Intersection of Jinzhou Road and Xinhua Road. The current house is an old residential property, the status the property rights belongs to public, under th unified management of the Heping District Housir Management Station. Surrounded by Xinhu Road Electronic Entertainment Street, where the gathering place of many young people, ju one block away from the Heping Road. With ve obvious location advantage, it can attract a lot people flow.

地块区位 Location

交通分析 Traffic Analysis

现状地块为一处老旧民居，在很小的空间内居住有 10 户，在建筑围合部分设有中庭，由于无人管理也缺乏有效的维护，中庭现在杂草丛生。在去场地调研后我们发现虽然整个中庭乱草丛生，这也成为我们之后将会主要考虑的设计元素。

The plot is an old local-style dwelling house and having 10 households in this small space. There is an atrium in the part of enclosed space, it is covered with grassed due to lack of effective maintenance and neglect now. After the research on the site, we find that although the whole atrium is covered with grasses, it also become the design elements we will mainly consider in future.

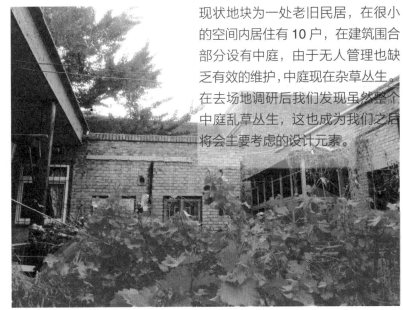

绿色中庭 Green Atrium

地块元素 Plot Element

从图上可以很明显地看出，由于缺少阳光，大部分居民的日常起居基本都是在中庭完成的。

Seen from the pictures clearly, most of the daily activities of the residents are done in the atrium due to lack of sunlight.

民居内有许许多多有趣的空间，这部剪刀梯就这一个狭小的空间内满足了交通的需求。

There are some interesting spaces in the dwellings. This scissors ladder meet the traffic need in this small space.

商业书吧 前期调研

等待拆迁的死气沉沉的社区

急切需要改善住房质量　　低收入生活质量不高

本地居民

当地政府　　开发投资　　有底蕴有文化的产业

盘活老城区发展　　　　　空间特点分明有吸引力

改变中心区城市形象

带来更多的财政收入　　廉价的土地和房屋使用费用

需求分析

在研究了周围的情况后，我们从地块内的各个利益方的需求出发，确定了我们设计的主要思路。本地居民最大的需求便是提高住房质量，尤其是冬天在没有暖气的情况下，居民日常起居都有很大困难。当地政府则主要希望盘活老城区的发展，由于临近和平路所以拆迁成本相当高，因而如何改善城市形象让政府很头疼。开发投资商最大的需求就是盈利，只有用低成本营造出有特色、有人气的商铺才能吸引投资商来注资，因此平衡三方利益也是设计的一个初衷。

After studying the situation around, starting from the need of all stakeholders within the plot, we confirmed the main idea. The greatest need of local residents is to change the o housing quality, especially in the case of no heating in winte people have great difficulty in daily life. Developer's greate demand is profit, only low cost to create distinctive stores ca attract investors to invest, so to balance tripartite needs is als one of our original design intentions.

街区内部缺乏有效的梳理和管理，存在许多乱停车、乱摆放的现象。内部交通流线也十分不畅通，同时建筑自遮挡过于严重，因此保留原建筑的方法可能无法实行。于是我们选择对原街区进行重新的梳理，将有限的地块作为街区改造的一个起点，来给整个街区提供更多的活动场地和公共空间。

For lack of effective planning and management, there are serious problems such as disorderly parking. Internal traffic flow is not smooth, and buildings are blocking sunlight gravely, so it is unrealistic to conserve the original buildings. Thus we plan for an overall renovation, starting with our limited plot. It will serve as a starting point to give the whole block more activity venues and public space.

树是一个街道的生命元素，在街区内部缺乏足够的绿色。沿街绿树成荫，整条街道在不拥堵时，步行感觉良好。

Greenery is an important element of a street which the inner block is lack. The street will feel cozy with trees lined up on both sides when there is no traffic jam.

街道上大量沿街空地利用困难，宽窄不一，很难成为通畅的人行通道。

The road is really difficult to use, because of the different width.

老旧街区乱停车现象极其严重，原本设计的车道被占用导致拥堵不堪。

The block is full of parking. The traffic jam is very serious.

虽然场地空间狭小，可是商业发展情况很好，沿街的绿植很多。

Though the space of this block is limited, there are many plants and stores here.

沈阳道古玩市场　　　　　　　　静园风景区

原地块料理店

沿线附近有相当活跃的消费潜力，如何与之区分开是定义自己店铺的重要因素。
There is consumption potentiality in this place. The important factor is how to figure out our difference.

周围诸如和平路、静园等一些旅游景点将带来一大部分游客。
There are many tourist attractions near us, which can bring many people to us.

地块附近的餐饮业十分发达，可以说从清真口味店到现代西式快餐店都有。
Catering services are popular here, which include many kinds.

业态统计分析

周围诸如和平路、静园等一些旅游景点将带来一大部分游客，因此将书吧与不同的业态相结合，以吸引旅游景点的游客。

Tourist destinations such as Heping road, jingyuan around bring a lot of visitors, so it is important for the bookstore to combine different kinds of commercial activities to attract tourists.

沿街面应尽量开放，内部地块则应该通过多样的空间丰富使用性。

The space along the street should be open as far as possible and we should rich usability through varied spaces.

中间内部设置中庭、保留地块原有的肌理也是我们考虑的因素。

They are our consideration elements to set an atrium in the central part and keep its original texture.

在靠内部的红色区块内应该布置工作室，并与中庭结合形成流畅的使用空间。

We should design a studio near the internal red part, which from a smooth service space combined with the atrium.

Comparing the weekly activity classifications of Douban.com, we find that photography is an important activity of the literary youth, rank closely behind socializing, so our focus is mainly photographers studio and public communication space.

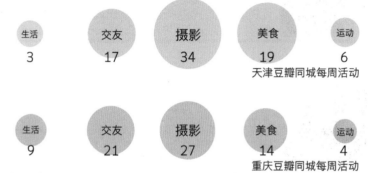

对比了豆瓣同城一周活动分类，我们发现摄影是文艺青年的一个重要活动类型，其次为交友，因此我们的功能设置便主要为摄影师工作室以及书吧的公共交流空间。同时我们也在比较中看出在北京一周的活动有上千个，因此天津的文化活动还有很大的发展潜力。

Meanwhile we also find there are over 1000 activities in a week in Beijing, so there is great developmental potential for cultural activity in Tianjin.

商业书吧前期调研

看到场地后,我特别想用一种特别开阔和轻盈的处理方式来缓解该地块给人的压抑感。同时会选用多元的组合方式进行组合,创造一个完全不同的空间。

Seeing the site, I would like to use a specially open and light way to remove the depression the site given. Meanwhile I will apply diverse combinations to create a completely different space.

我自己特别喜欢写作,所以我特别想把这里变成能够进行创作、听取不同人的故事、类似于茶馆的空间,因此我更多地从功能出发来考虑这个设计。

I personally like writing, so I want to turn this into a place where people can write and listen to the story of others, something similar to a teahouse. So I will start from this function to conceive the design.

该地块的最大特色我觉得是老旧街区赋予的。因此在街区里以一种完全不同的手法改变肌理是不可取的,小型的植入与填充的改造应该更加适合。

The feature of the site is endowed by the old block. So it is undesirable to change the texture completely, while slight implantation and modification should be more suitable.

作为一个天津人,我对于沈阳道古玩市场等地方是有感情的。这些老旧记忆中留存有很多工艺和材质,所以如何让书吧与众不同,我的选择是砖和木。

Born in Tianjin, I feel nostalgic for the antique market on the Shengyang Road. These memory comes along workmanship and materials, so I try to make the book-cafe unique with brick and wood.

南向树木

建筑间隙小

功能分析讨论
Functional Analysis Discussion

建筑自遮挡

在确定功能上，我了解到现在中国忧郁症的发病率极高，书吧或许可以缓解此问题。

The onset of depression being extremely high in China, it comes to me that book-cafe may help to relieve.

我对于功能的设定并不复杂，就是创造出讲故事的空间给作家提供机会。

Function of my book-cafe is not complicated: providing chances for writers by creating space for story-telling.

我主要是想做一个非常纯粹的书的空间，利用廊道和各种新产品来达到最佳的读书体验。

I want to create aspace purely for reading, use the corridor and new products to achieve the best reading experience.

木头和砖石是我这次对于记忆的还原点，因此能够将木艺融入空间里是我的追求。

Wood and brick are origins of memory, so incorporating wood craft into space is my pursuit.

功能分析讨论
Functional Analysis Discussion

围墙遮挡商铺

建筑入口狭窄

空间分析讨论
Spatial Analysis Discussion

建筑自遮挡

窄路小巷模式

原来的老旧街区建筑自遮挡非常严重，因此设计的首要任务是如何创造更多的明亮空间。

窄路密网下绿色植物显得尤为重要，在这个地块内任何建筑都必须解决和绿植的关系。

原有旧民居有许多富有特色的空间，例如中庭和架空廊道等，保留这些将提供更多的空间层次。

中庭完全露天，在使用上会带来很多问题，设计一个室内的中庭也是一种考虑。

As the blocking of sunlight is quite severe, the priority should be how to create a spacious space.

Greenery is especially important in the narrow alley, so buildings have to be in harmony with plants.

There are many characteristic spaces like yard and overhead gallery, which can help to create more levels.

A yard completely exposed is questionable, but an indoor yard can be a good option.

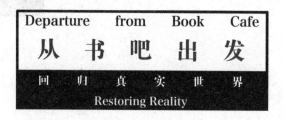

— 体验你的书房 —

格儿书行
CER BOOKS

提供富有情感的职能书房解决方案

吴嘉琦
Wu Jiaqi
"文艺中老年"

Commercial Plan

前期商业策划

如今整个社会已经进入了互联网时代，尤其在商业上，已经呈现出电商独霸的势头。但同时电商最大的短板——用户体验也逐渐在电商激烈的竞争中越发被重视，许多品牌也都在线下设立了自己的体验场所。然而大部分体验店的效果并不理想，在我们网络调查和现场实地调研后，发现原因主要有以下几个方面：
1. 品牌单一，体验店影响力不够；
2. 大型商场租金过高；
3. 体验店设计不够人性化，用户体验感不佳。

Nowadays, the society has been becoming the Online Era, especially in the business. It has been occupied by the Giant of e-commerce. And their biggest shortcoming is becoming more important to them. A lot of brands built many experience shops, but the result is in vain. After our survey with many different aspects, we find some reasons:
1. Only brand's sole experience shops can't be very in fluential.
2. The rent of big mall is too expensive.
3. The design of experience shops is not humanized, the feeling of experience is not good.

PLAN SCHEME AND MARKET BACKGROUND INTRODUCTION
策划方案及市场背景介绍

在O2O市场，完成了从线上转移到移动终端的战略后，越来越多的电商平台开始转战线下体验服务，因此这也将是一个巨大商机。

In the O2O market it completed the shift from online to mobile terminal strategy, more and more E-business platform began to experience offline service platform, so it will be a huge business opportunity.

电商的竞争
Competion of E-business

冷清的传统商业
Depression of traditional business

活跃的线下体验店
Popular offline showrooms

寻找新的营销模式
Search for new marketing modes

产业大变革时代
Time for industry revolution

② PLAN SCHEME AND MARKET BACKGROUND INTRODUCTION
商业创意和营销活力

从一本书究竟能成为怎样的商机开始思考，对不同的看书需求进行分析，通过咖啡、电子产品、家具、灯具等，营造出浓郁的书房氛围。

From beginning think how a book become a business opportunity to analyze the different needs for books, to create a strong study atmosphere through coffee, electronic products, furniture, lamps and lanterns.

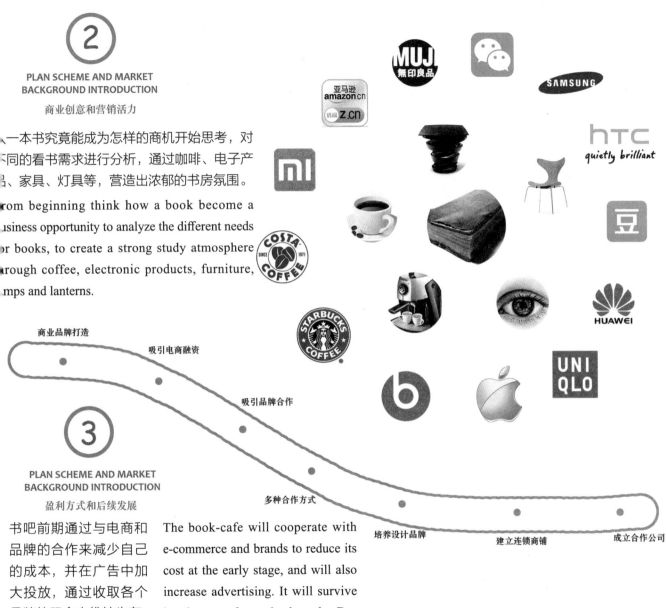

③ PLAN SCHEME AND MARKET BACKGROUND INTRODUCTION
盈利方式和后续发展

书吧前期通过与电商和品牌的合作来减少自己的成本，并在广告中加大投放，通过收取各个品牌的租金来维持生存。后期通过培养自己下属的设计品牌，利用自己打造的平台盈利。

The book-cafe will cooperate with e-commerce and brands to reduce its cost at the early stage, and will also increase advertising. It will survive by the rent from the brands. But after that, the book-cafe will foster some own design brands, taking the advantage of its terrace to get profit.

Ger Books

格儿书行

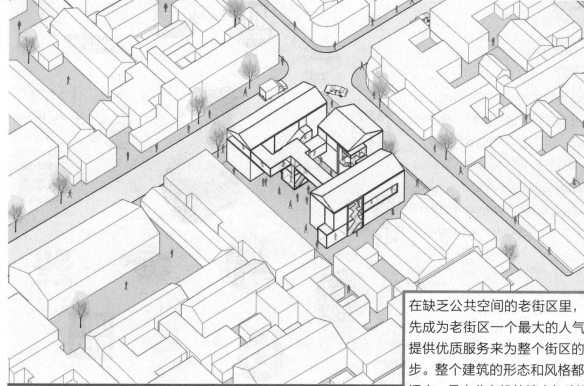

在缺乏公共空间的老街区里,格儿书行将首先成为老街区一个最大的人气带动点,通过提供优质服务来为整个街区的更新迈出第一步。整个建筑的形态和风格都与周围现状相适应,是十分有机的填充与改造。

In the old block lacking public space, Ger Books will be the biggest beauty spot, bringing the best service to the block. The form and style of the book-cafe is really adapted to the block. And it is very organic dunnage.

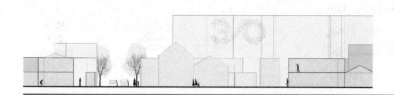

街道剖面图

空间的概念上，书吧利用不同的廊道给人一种能自由通行的感受，同时不同的廊道设置在不同的建筑位置，由交通盒进行区分，三层共设有八个不同类型的廊道,根据位置的不同进行了细致的区分。一层廊道较为开放，为可以自由移动的空间；二层廊道较为私密，设有可以坐的小型空间；三层廊道最为私密，设置有相当多的半封闭的阅读区。

For the space concept of the book-cafe, it uses different galleries to make people feel they can walk inside freely. And the galleries are around everywhere. Different floors, different position and different aspects have different functions. And all the galleries are using the box of communication to link to each other. The book-cafe has 3 floors and 8 galleries for used. The galleries on the first floor is open, people can move freely. The galleries on the second floor is a little private where there are some small spaces people can sit. The galleries on the third floor is the most private where there are quite a lot of semi-closed reading area.

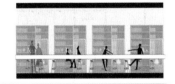
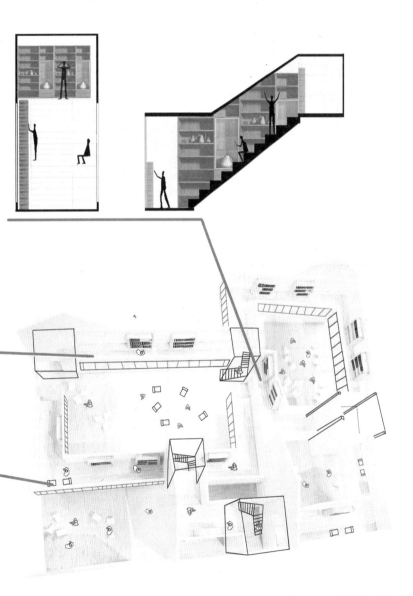

模型表现图

平面图

在平面布置时，三个主要体块的建筑依据距街道的距离分别设计为售商业的饮食区、休闲咖啡区和办区。二层及二层以上均为书廊和空筒空间，主要活动大空间均设在一以方便使用。同时交通筒的设置是平面更有趣，流线多变，空间层次明。

As for layout, the three main blocks building are designed as commerce food area, leisure coffee area and of area according to the distance betw the block and the street. Above f floor are the "book corridor" and "tra cone", which create various route a clear division.

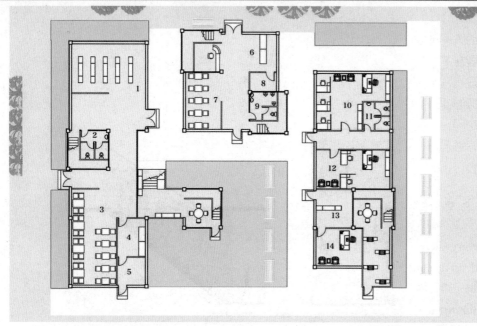

一层平面图 1：150

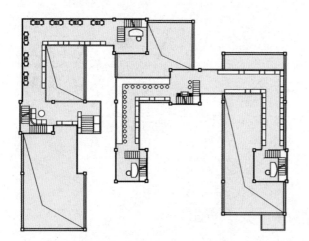

二层平面图 1：200

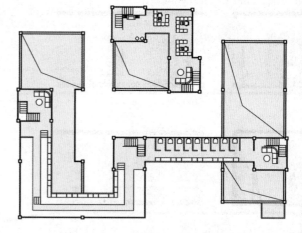

三层平面图 1：200

1. 设计师品牌
2. 女卫生间
3. 西餐厅
4. 厨房
5. 仓库储物间
6. 咖啡吧柜
7. 咖啡隔间
8. 操作间储物间
9. 男卫生间
10. 出版社办公室
11. 卫生间
12. 书吧办公室
13. 摄影师展厅
14. 摄影师工作室

剖透效果

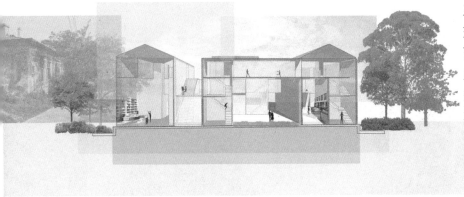

三个主要的坡屋顶大空间将各个交通盒包含在其中，同时解决了两个重要的问题：一是内部廊道系统错综复杂，但缺乏主要的大空间；二是老旧街区不宜出现极其突兀的异形建筑，因此选用原有的材质和手法将方案较为完整地表现出来。

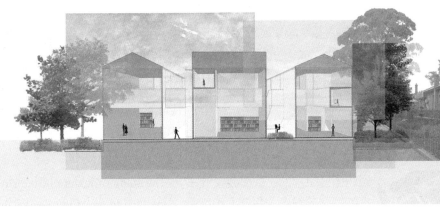

Three main big rooms with slope roof include all the traffic box. It solves two main problems at the same time. The first is the complex galleries which make the building be lack of the major space. The second is the old block. Our building must be adapted to the block and its age. So we choose original materials and method to show the project completely.

交通盒

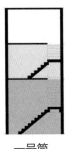
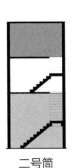
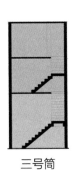
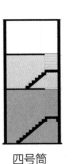
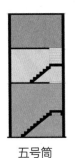

一号筒　　二号筒　　三号筒　　四号筒　　五号筒

自动售货　接待咨询
休闲娱乐　厕所卫生
书籍查询　讨论交友

SAVING OR ABANDONING

明亮的墙体 BRICK WALL
斜屋顶 SLOPING ROOF
非法结构 UNLEGAL STRUCT
插入悬空走廊 PLUG VACANT CORRIDOR
复古装饰 RETRO DECORATION
混乱的电线 CHAOTIC WIRE
缺少阳光 LACK OF SUNLIGHT
混乱的电线 CHAOTIC WIRE
舒适的小巷 CONFORTABLE ALLEYWAY

图底关系

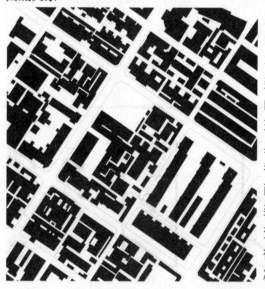

对于街道肌理的研究是我设计书吧的一个出发点，在老旧街区里不应该出现一个异形或者奇怪的建筑，同时利用地块内原有建筑的各种优点和优秀的手法进行处理，包括坡屋顶、架空廊道、古典的装饰和高宽比为3：1的小巷空间。同时要将这个街区沿街面进行治理，保留原有的砖墙铺装，让这个本来缺乏生机的地块变得更加有趣起来。这样在保留街区味道的同时又在一定程度上减少了成本，另一方面也让特色变得更加鲜明。

The study of texture of the street i the beginning of my design. In ol block, it is unsuitable to show up a alien or a strange building. I mak use of the advantages of the bloc to design and solve the problems including slope roof, galleries classic decoration and the space o alley. Also to finish a great design I have to finish transformation o the street. After that, this area wil be more interesting and stay its ow style. And the cost of all the projec will be cut down to a very low level

| 格 儿 書 行
GER BOOKS

也块选择在新华路和锦州道的交口处，这里靠丘和平路恒隆广场和滨江道步行街。但由于基出设施较差，人员结构复杂，加之建筑年久失多，导致价格很低，这也是我最后选择这块地的原因。同时在房屋土地产权上做了浅度的调研和分析后，发现了租借地块老旧城区容易出现产权混乱的问题。同时在商业模式策划这个内容里，更好地学习和领悟了设计的内涵。任何一个设计一定要经过不断的返工、磨合，最终与真实世界的需求完美地结合在一起，形成螺旋上升，而这也正是我们在上设计课时在大平图后就将整个设计抛在脑后的弊端所在。

We choose the site in the cross of the Xinhua road and Jinzhou road, near the Hanglung Plaza and the Binjiang avenue pedestrian street. But because of the poor infrastructure, the complex personal structure and disrepair construction, the prices of this place are very low. That's why we finally choose this plot. We also did some research on housing land property, found the plot in old town usually has the chaos of loan property. At the same time, we learnt and understood the connotation of design. Any deign will be going through a rework running-in, the connect with the real world needs.

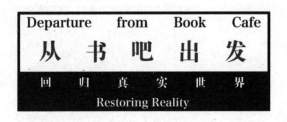

LONELY BOOKTALK
春田花花谈书会
PSYCH-BOOK

人生中总有那么<u>一本</u>你想**看**
人生中总有那么<u>一刻</u>你想**说**

王卓
Wang Zhuo
"只能我放弃人生,不能人生放弃我。"

Psychology Consulting 心理咨询服务

Books&Cafe 书与咖啡

优质图书 Selected books

深夜提供低度数酒精饮料
Non-alcohol drinks

吸烟区与非吸烟区
Smoking area and non-smoking area

舒适的阅读场所
Comfortable reading atmosphere

WHY 把二者融合在一起?

书店的氛围能够缓解、稳定紧张情绪,让心理焦虑、抑郁症患者在不知不觉中获得治疗

The book store has a soothing effect for depressions

Books&Psychology 书与心理

HOW TO 把二者融合在一起?

除了独立分区提供私密心理咨询外,书店内阅览区与咖啡区也提供有隔断的空间提供心理咨询服务

Beside private counselling area, there are also separate places in reading area and cafe that provides counselling

两者关系

图书市场　心理市场

为心理市场提供场所、塑造气氛,为心理市场提供周边产品(心理相关书籍、商品)。在除了与心理市场相互作用以外,也拥有自己独立的客群

为图书市场供客源,成书店经营的头之一,作高利润行业图书的经营供盈利保障除了与图书场相互作用外,拥有自独立的客源目标市场(业服务)

Book markets
Provides places for psychological markets, a provides accessories for the psychological market has its own targeted customers too.

Psychological markets
Provides customers for the bookstore, and becom a gimmick for the marketing of bookstore in order obtain higher profits

宏观政策
Policy

国家"十一五"规划纲要
"六个立足"中"以人为本"
增强国民的幸福感、心理
卫生指数
Human-oriented planning, enhancing the inner happiness of citizens

市场与消费群
Targeted customers

高校、企业与中高收入家庭；
中高收入的受教育人群、青少年、老年人
University students, middle-high income families, educated people, adolescents, elderly people

产品与服务
Products and services

心理咨询与心理治疗；
个人心理测量；
企业员工心理培训；
企业咨询；
大众心理课堂
Psychological counselling
Personal psychological management
Enterprise psychological training
Enterprise counselling
Open courses

方法
Methods

（咨询室）面谈法；
（阅览区）面谈法；
线上咨询
Face-to-face counselling (private)
Face-to-face counselling (public)
Online counselling

竞争 Competition

医院心理门诊；Hospital
价格较低，一般为 50 元 / 小时；
氛围使人紧张，偏重于药物引导和治疗
Lower price, 50 yuan per hour. Focusing more on pharmacotherapy

学校心理咨询；School counselling
免费，由教育部规定设立；

专业性稍弱，消费者担忧隐私泄露等
Free, regulated by the education department of the government. Less professional, customers are concerned with privacy divulge

社会民营心理咨询；Individual therapist
价格较高，一般为 200+ 元 / 小时；
没有市场细分，针对性不强
Higher price, 200+yuan per hour. Lack of market segmentation.

合作 Cooperation

高校合作 University
与天大、南开等高校合作，共同研发或提供咨询优惠
cooperate with tianjin university and nankai university, provide counselling coupons for students

企业合作 Enterprise
与部分企业合作，提供企业培训合作公益事业进行宣传
cooperate with enterprises, provide training and open classes

两大模式共同推进
书店与心理服务共同经营

心理服务项目名称	第一年（万元）
服务设备费	8
办公设备费	8
开办费	2
人员工资	22

心理服务项目名称	第一年（万元）
心理咨询与心理治疗	30
心理测量	5
企业服务	70
其他收入	10
心理服务收益	115

书店项目名称	第一年（万元）
装修	30
办公设备费	2
人员工资	50
前期书籍/商品投入	20
一期总投资	约230

书店项目名称	第一年（万元）
书籍销售	100
商品	30
咖啡店	50
书籍销售收益	180

共同项目名称	第一年（万元）
租金投入	90
市场推广与宣传	1
管理费用	2

共同项目名称	第一年（万元）
政府扶持	150
长期借款	50
长期可支配现金：投资方投入	50

其他修改

书店与阅览室外立面更加通透，夜晚更加迷人的灯光。

阅览区内部分隔更加多元，有相对私密区域，增加隔断。

阅览区增加对院子开放的程度，更加接近自然。

简化室外阅览区繁杂的小尺度的设计，向垂直花园方向改进。

改进体块，在保留原有街道肌理的情况下，解决咖啡馆与街角流线问题。

本项目通过研究历史街区图底关系，对该块城市肌理的基本尺度掌握以后，在不坏肌理的同时对建筑形式进行创新。

咖啡区、办公、书店、阅读等功能分区分以长方体为基本型，看似分离，客观上过廊道连接，心理上通过共享庭院等公空间连接。

建筑设计阶段就初步考虑策划与营销的法，提出以"你好，忧愁"作为室外阅空间设计的基本概念。

By studying figure-ground of the historic district and getting a basic comprehension of the scale of arrangement of the block, this project is designed to protect the existing fabric while making innovation in the form of building.

By zoning the site into cafeteria office, bookstore and reading area, this project connects areas of different functions physically by corridors as well as mentally by sharing public courtyard.

When designing the building itself, some ideas of marketing have been adopted. The notion of "hello sadness" is the leading concept when designing

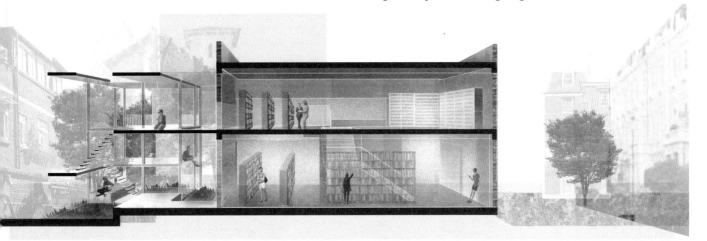

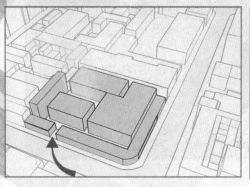 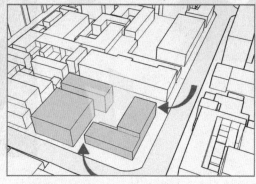 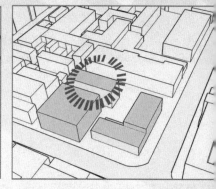

原有建筑 Existing building	新建筑 New building	体块局部 Form
底层商铺 Shops on street level	延续肌理 Maintain city texture	红圈体块 Highlighted space
完全围合 Fully closure	半围合 Half closure	方便进入 Easy access
一个入口 Single entrance	两个围合 Two entrances	相对私密 Privacy
庭院狭窄 Cramped spaces	庭院多功能 Multi-use courtyard	采光较好 Better lighting

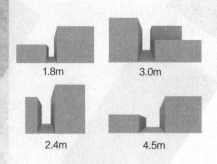

40 × 15

庭院尺寸

提升布局（整理流线 / 成为来自不同功能区人流的交会点）
Better layout

采光（光量重新调整）
Lighting

改善环境（围合幽静，种树）
Enhance environment

聚集人气（成为最具有人气的部分，空间分隔 / 灵活空间配置）
Gather people

传承（原本住宅中院的形态传承）
Inheritance

分区分析 Division analysis
沿街立面，吸引客源
商业性功能布置在沿街
Open facade for attracting customers
Retail on the street level

特色流线 In-and-out
对内对外
二层连接
A round circulation on both
first floor and second floor

特色场所 Special space
街区现代垂直花园
合尺度的室外阅览
Vertical garden in the traditional scale
Open reading space

特色庭院 Special courtyard
举办活动
聚集人群
Hold activities
Gather people

一层平面图 1st Floor Plan

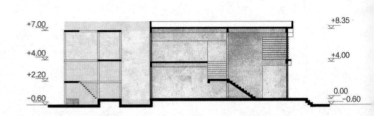

剖面图 A-A   Section A-A

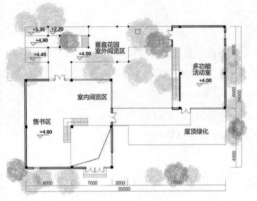

二层平面图 2nd Floor Plan

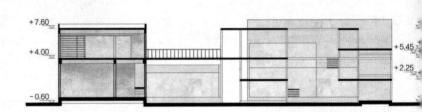

剖面图 B-B   Section B-B

很多东西是要传达到别人那里才算是概念的完成。我做了不计其数的展示，以及不计其数的演示文档和主题发言，甚至剪了一个蒙太奇的视频，再加上终期之前无数次的排练，要表达清楚一个简单的概念，是需要完整的逻辑的。

A lot of things able to be transfered to others is the accomplishment of concept. I had made countless exhibitions, PPTs and keynotes, even cut a montage video, plus many rehearsals before the dealline, it needs a complete logic to express a simple concept clearly.

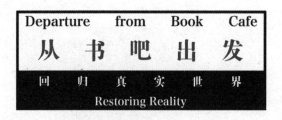

冲冲手工书坊
Chong's Handcraft Bookshop

我的书，你的木

刘冲
Liu Chong

"念念不忘，必有回响。"

文艺的人是谁
Who is the poet ?

有思想　　善于感动　　活得精致　　善于发现美
Minded　Sensitivie　　Fine　　Esthetic

爱读好书　　懂点艺术　　通晓琴棋书画　　心灵手巧
Literature　　Art　　Sketch　　Skill

——我们"贩卖"体验
Selling experienc

阅读体验

Reading Experience

通过与纸质书籍的接触,享受在书海中的偶遇,是刨除畅销书和快餐书籍的精品阅读。

Feeling the texture of paper, enjoying the plot of piles of books and eliminate fast-reading and best-sellers.

空间体验

Space Experience

在专门营造的空间中,感受独特的光影,体验过去与现在的融合,以及高雅宁静的氛围。

In specially designed space, feel the interplay of light and shadow and the fusion of present and past.

休闲体验

Entertainment Experience

这里有咖啡、茶点解乏,有设计精美的工艺品,有投你所好的沙龙讲座,甚至教授工艺的课堂。一切都来源于生活。

There are coffee and tea, delicate handcrafts, salons and DIY classes, which exemplify the esthetics of life.

文艺的人爱什么？
What does he love？

木作
创意产品
以物易物
书籍

Woodworks
Creative Product & Furniture
Goods Swap
Books

木作课堂上亲手制作木艺作品，完成设计、划线、锯切、表面处[理]等所有环节，对优秀作品进行展示与投产。
In woodcraft classes, you can design and process woodworks with th[e] help of master. We display and sell excellent works.

这里是一个以书为线索的生活体验馆，接触到的家具、摆件都是[我]们的木作产品，顾客可以把在这里收获的部分感动带回家。
The studio is an experience pavilion, Furniture and decorations here a[re] all our hand-made products, which can be bought home.

以等值书或木制品交换独一无二的物品，手绘明信片、旧书、手[工]艺品，投产顾客的创意产品，重新利用旧物做成创意品。
Exchange unique goods with books and wood crafts, sell student[s'] products, turn old things into creative products.

——我们"贩卖"情[怀]
Selling feeling

商业模式
Commercial Model

核心主业链 Industry Chain

自主设计创意品
Independent Design

线上征集方案，
鼓励网络传播，
提高影响力。
Collect design projects online,
Encourage sharing on internet,
Enhance popularity.

样品制作
Sample Manufacture

会员参与制作，
提供体验服务，
节约成本。
Customer maufacture,
Experience service,
Save cost.

众筹生产
Crowd Funding Produce

线上众筹预定，
人数足够后投入生产，
快速规模化。
Crowd funding online,
Product with enough customer,
Grow into large-scale.

盈利模式 Profit

收 INCOME:

书籍售卖	Books
文创产品售卖	Cultural & Creative products
家居用品售卖	Furniture & Household items
咖啡店售卖	Café
场地租赁	Venue rental
课程报名	Course registration
网上商城收入	Online shops

支 EXPENSES:

土地租金/合营出版社分担 10%	Land rent/10% shared by the joint press
装修成本	Renovation cost
书籍成本/出版社 6.5 折提供	Books cost/35% off provided by the press
员工成本	Staff wages
家具成本/季度更新自主设计，40 万/季度	Furniture cost/seasonly designed, 0.4 million each season
网络运营费	Network operation cost
品牌推广费	Brand promotion cost

建筑设计
Architecture Design

书吧的设计从"感受生活之美"的品牌文化出发,保留原有老建筑的桁架和红砖,围合出一个"老"空间供交流和活动之用,其他功能围绕其展开,以材质对比强化观感。沿街两侧排布咖啡区以及文创产品销售区,最大限度地利用沿街界面。将书籍阅览区放在内向区域。二层中间挑空,顶部开窗,突出"老"空间的存在。西南立面作为光的主要来源方向,通过砖的堆叠变化,营造出静谧、神圣的光影效果,烘托书吧内的阅读氛围。

Featured by *Feeling the beauty of life*, the book-cafe constructed with the original old building trusses and brick which enclose an Old space for communication and activiti with other functions around it. The window in the ceilir highlight the existence of the Old space. Southwest facade the main direction of the light source, the change by stacki bricks, create a tranquil and sacred lighting effects, enhanci the reading experience.

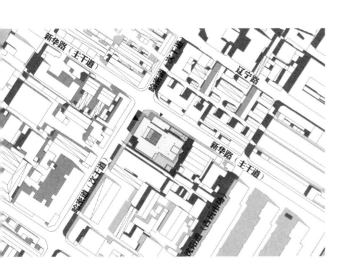

以中心"老"空间为核心，对四周的空间产生一定的引力，
与之发生空间上的联系。二层的挑空使空间整体上更完
整、更通畅。在整个空间内，"老"空间呈现出一种展品的
姿态。

he "old" room in the center as core, generate a certain
ravity and contact with the space around. Soaring space on
e second floor as a whole make the space more complete,
ore unobstructed. In this space, "old" space shows exhibition
esture.

方案中最为核心的部分就是这个"老"空间。它保留了街区内原有的砖墙和桁架，其他的功能也是围绕它来㊥布的。与西南面用砖块随机排布形成的立面相呼应，祂它们之间形成一种微妙的空间。

The most central part of the program is the "old" space. retains the original brick wall and trusses, the other function are arranged around it. Facade on the southwest with random brick arrangement echoes with the old space, ar form a subtle space between them.

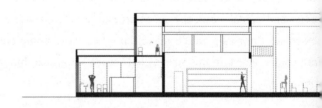

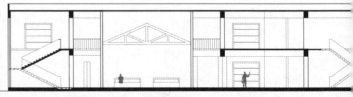

感受生活之美
Feeling the beauty of life

身于老房子前，回念往事，感受此刻的静谧，憧憬未来美好。

样的设计所希冀的就是用街区自己的东西来唤起人们的忆与共鸣，让这个书吧参与到人们的生活中去，用空间营造的情景打动人，进而激发人们内心对生活的热爱。

Standing in front of the old house, we recall the past feel the quiet moment and look forward to bright future.

Such a design aims at using blocks' own things to evoke memories and resonance, so that the book cafe will participate in people's lives, create a scene to impress people, and then guide people's inner love to their life.

后记
Postscript

本书工作自 2015 年底开始,历时近两年。作为一次探索性的教学方法和设计的研究,本项工作得到了天津大学建筑学院领导和城乡规划系运迎霞系主任、许熙巍副系主任等的大力支持,特别感谢城乡规划系本科二年级教学组的姜耀明、李泽、邢锡芳等老师的鼓励与支持。

本书工作由左进副教授全面负责,同时邀请了郭泰宏(天津九界投资管理有限公司策划副总监)、齐晓茜(天津九界投资管理有限公司高级策划师)、李晨(天津市城市规划设计研究院规划师)等进行指导与交流,参与的同学包括:吴嘉琦、王冠仪、牟彤、刘冲、王卓、蒋瑞、汪梦媛、林玉、杨力行、林澳。在这个团队里,每个参与者都在互相学习和积极贡献中扮演着不可或缺的角色。所以,本书是集体智慧的结晶。

在本书的出版过程中,吴嘉琦、王冠仪为内容编辑提供了全面的支持,陈昱安为本书设计了封面,本书策划编辑陈景女士为本书的出版付出了艰辛劳动。特别感谢虎尾科技大学休闲游憩系教授兼文理学院院长侯锦雄教授为本书作序,侯老师的鼓励与支持将激励我们继续努力奋进。在此,我们对所有参与和支持本书工作的人士表示衷心的感谢。

在本书即将付印之际，感慨良多。在多年的教学、研究与实践工作中，我们一直坚持以真实的城市空间为研究对象，探究其面临的真实问题。在本次尝试中，我们希望以"书吧设计"为出发点，立足于真实的城市区域，用天真的眼睛观察世界，运用互联网平台与田野调查等，了解使用者（经营者和消费者）需求，剖析现实问题，探讨书吧"想在哪里""想成为什么""想怎样活着"，尝试在以空间设计为主的设计教学中做出一些探索和突破，整合"市场分析、场地调研、项目策划、空间设计"等四大板块，从不同维度思考并为城市书吧提供一体化的解决方案。

本书的完成，是一个不断思考、求教、探索的过程，限于时间与经验的不足，余谨以虔诚之心，乞教于学界业界的师长和朋友。最后，谨以费孝通先生在《乡土中国》书中的一段话与大家共勉：

"我并不认为教师的任务是在传授已有的知识，这些学生们自己可以从书本上学习，而主要是在引导学生敢于向未知的领域进军。作为教师的人就得带个头。至于攻关的结果是否获得了可靠的知识，那是另一个问题。"

左 进

2017 年 8 月于天津大学

图书在版编目（CIP）数据

从书吧出发：回归真实世界 / 左进著． -- 南京：江苏凤凰文艺出版社，2017.11
 ISBN 978-7-5594-0687-3

Ⅰ．①从… Ⅱ．①左… Ⅲ．①书店－建筑设计 Ⅳ．① TU247.9

中国版本图书馆 CIP 数据核字 (2017) 第 133571 号

书　　　名	从书吧出发——回归真实世界
著　　　者	左　进
责 任 编 辑	聂　斌　孙金荣
特 约 编 辑	陈丽新
项 目 策 划	凤凰空间/陈　景
出 版 发 行	江苏凤凰文艺出版社
出版社地址	南京市中央路165号，邮编：210009
出版社网址	http://www.jswenyi.com
印　　　刷	北京博海升彩色印刷有限公司
开　　　本	889毫米×1 194毫米　1/16
印　　　张	6.25
字　　　数	80千字
版　　　次	2017年11月第1版　2023年3月第2次印刷
标 准 书 号	ISBN 978-7-5594-0687-3
定　　　价	68.00元

（江苏凤凰文艺版图书凡印刷、装订错误可随时向承印厂调换）